**NISTIR 7517**

# Extending the Notion of Quality from Physical Metrology to Information and Sustainability

Gaurav Ameta
Sudarsan Rachuri
Xenia Fiorentini
Mahesh Mani
Steven J. Fenves
Kevin W. Lyons
Ram D. Sriram
*Manufacturing Systems Integration Division*
*Manufacturing Engineering Laboratory*

July 2008

U.S. Department of Commerce
*Carlos M. Gutierrez, Secretary*

National Institute of Standards and Technology
*James M. Turner, Deputy Director*

## Table of Contents

Abstract .............................................................................................................................. 1
1     Introduction ................................................................................................................ 2
2     Physical Quality ......................................................................................................... 3
3     Information Quality ................................................................................................... 7
4     Sustainability Quality .............................................................................................. 12
      Sustainability metrics ............................................................................................... 18
      Future directions ...................................................................................................... 20
5     Discussion and Comments ....................................................................................... 21
References ......................................................................................................................... 26

## List of Figures
Figure 1. The quality triangle as used in this paper ...................................................... 3
Figure 2. Logical relationship among metrology concepts for use in the standardization in measurements (from [10]) ........................................................................ 4
Figure 3. Classification of tolerance types as defined in the ASME Y14.5 standard. 6
Figure 4: IQ dimensions mapping ( [26] , [5]) ............................................................ 11
Figure 5. Introduction of the sustainability aspect for deciding the function of a product. ........................................................................................................................ 14
Figure 6. Possible system for producing sustainable products .................................. 15
Figure 7. Potential connections between the criteria for measuring the quality of sustainable products and the environmental impacts from EPA [63]. ...................... 16
Figure 8. Framework relating information representation and Product Life Cycle (from [52]) .................................................................................................................. 22
Figure 9. Triple bottom line of sustainability being viewed from a lens of influence of technology. ............................................................................................................. 23

## List of Tables
Table 1. Summary of standards for sustainability ...................................................... 18
Table 2. Comparison of current state of Quality in Physical, Information and Sustainable realms. ..................................................................................................... 24

# Extending the notion of quality from physical metrology to information and sustainability

Gaurav Ameta, Sudarsan Rachuri, Xenia Fiorentini, Mahesh Mani, Steven J Fenves, Kevin W. Lyons, Ram D Sriram

**Abstract**

*In this paper we intend to demonstrate the need for extending the notion of quality from the physical domain to information and, more comprehensively, to sustainability. In physical metrology there are well established principles such as fundamental units, precision, accuracy, traceability and uncertainty. In order to understand and define quality for information and sustainability we need to develop metrological concepts similar to those of physical metrology. Research efforts related to information quality (IQ) are scattered. IQ is primarily defined in terms of several characteristics (dimensions) which lack consensus definitions and are sometimes subjective. However, the notion of IQ currently in practice has provided some useful insights towards defining formal approaches to IQ.*

*In order to extend the notion of quality to sustainability we need, as in the case of information, a well defined metrology similar to physical metrology. Sustainability is currently getting attention in many areas of human endeavor. One proposal is to measure sustainability in terms of a triple bottom line, namely social, economical and environmental aspects of human endeavor. Sustainability metrics are continuously evolving and their clear definition is fundamental to the understanding of the notion of sustainability quality. After analyzing the current literature, we identify the following needs for characterizing the notion of sustainability quality: a) standardized terminology of terms and concepts, b) metrics and metrology, c) harmonization and extension of standards, d) conformance testbeds for standards and e) development of information models that support sustainability.*

Keywords: quality, metrology, sustainability, information quality.

# 1 INTRODUCTION

Enterprises compete by offering better quality in their products and services. Quality, as defined in the survey by Hoyer and Hoyer [1], has two levels. At the first level, quality refers to producing products or delivering services with measurable characteristics satisfying a fixed set of specifications that are usually numerically defined (objective definition). At the second level, quality is about producing products or delivering services that satisfy customer expectations for their use or consumption (subjective definition). The ISO 9000 standard defines quality as the "degree to which a set of inherent characteristics fulfill requirements" [2]. The American Society for Quality defines quality as "a subjective term for which each person or sector has its own definition" [3].

Most definitions of quality indicate the satisfaction of both subjective and objective requirements. Objective quality can be measured using metrics identified in relevant standards (if available), while subjective quality, due to its inherent nature, is difficult to measure. For this reason in this paper we only consider the objective quality of products.

Generally, the quality of products is associated with physical quality, but in our research towards sustainable manufacturing we extend quality to include information and sustainability. We present the quality of products in terms of three attributes: physical, information and sustainability. Figure 1 illustrates the notion of quality as used in this paper.

Physical quality corresponds to all physical properties, such as geometrical, mechanical, and material properties. Information quality corresponds to all product/ process information traceable throughout the product lifecycle and involved processes. Sustainability quality corresponds to the environmental, societal and economical impacts throughout the product's lifecycle.

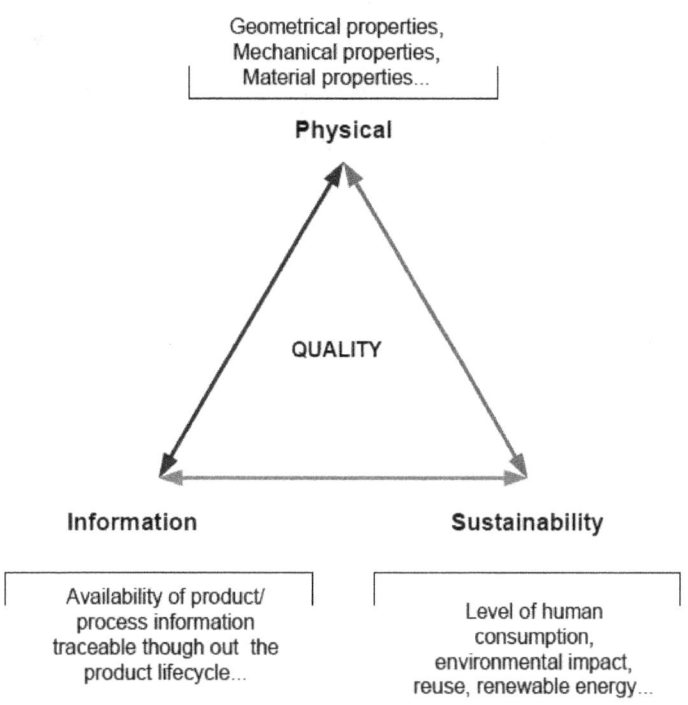

**Figure 1. The quality triangle as used in this paper**

From the figure, it can be seen that these quality attributes are mutually dependent, thereby making product/process quality a composition of the three attributes. While the physical quality of products has been well studied and quantified, there have been only limited efforts aimed at understanding and quantifying the information quality of products [4; 5] and only exploratory efforts have been conducted on understanding the sustainability quality [6; 7].

Subsequent sections of the paper discuss these three attributes in greater detail.

## 2 PHYSICAL QUALITY

Well established principles, namely, fundamental units, precision, accuracy, traceability and uncertainty, govern physical metrology (the science of measurement). Measurement in general is the process of quantitatively comparing a variable characteristic, property, or attribute of a substance, object or system to some norm [8]. For example, the goal of the International System of Units is to define a system of measurement units (norm) directly based on quantities of nature that do not vary with time or circumstance (e.g., the meter – the unit of length - is defined in terms of the speed of light). The measurement process is performed using appropriate instruments. To certify an instrument's accuracy relative to a known standard, a mechanism called traceability is used. This mechanism is intended to assure an unbroken chain of measurements, each having stated uncertainties, relating a given instrument's measurements to a known standard [9].

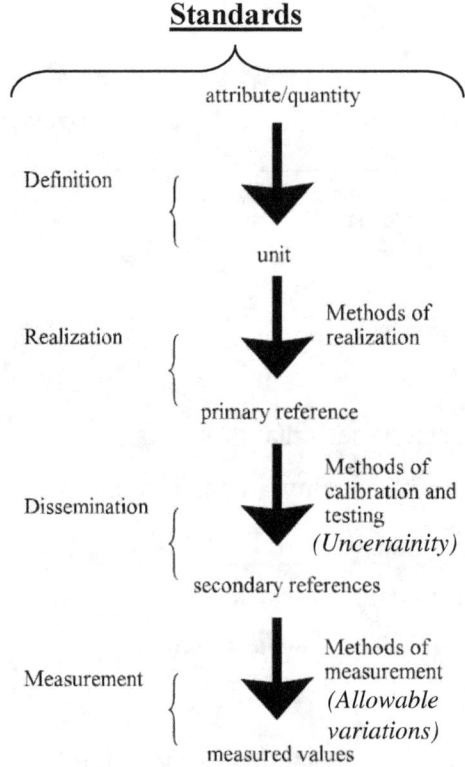

Figure 2. Logical relationship among metrology concepts for use in the standardization in measurements (from [10])

Figure 2 presents the logical relationship among metrology concepts as presented in [10]. It demonstrates that standards define units for an attribute/quantity, create methods of realizing the attribute/quantity (through an instrument), then create methods of calibrating and testing the instrument and finally develop methods of measurements. Each instrument has a particular precision and therefore generates associated uncertainties in the measurements. Keeping this in mind, allowable variations in the measured quantity also need to be specified. For more information on the standardized terminology for physical metrology, refer to [11; 12].

Measurements are performed on products to ascertain their physical quality. The physical or mechanical quality of a product includes specification (geometric, material, etc.) at the design stage, process control at the production stage, and inspection for adherence to the design specification and testing of the final product. For each quality requirement there are standards to follow and then metrics to measure the variation from the standard. Since products cannot be manufactured to exact specifications, there must be allowable variations in the values of the measured metric from the intended value.

An example is geometric tolerancing, the modern method of specifying allowable limits in the geometry of product. There are two similar standards in use for specifying tolerances, ISO 1101 and ANSI/ASME Y14.5M [13; 14]. These standards have classified dimensional variations (size) and geometric variations (form, orientation, profile, position, and runout) into separate classes (Figure 3). As shown in the figure, form variations are subdivided into: straightness - applicable to line segments (e.g., axis), flatness - applicable to planar geometry, circularity – applicable to circular geometry and cylindricity – applicable to cylindrical geometry. Similarly, other geometric variations (orientation, location, runout and profile) are also subdivided based on the applicable feature, its reference (datum) and other such requirements.

Depending on the functional and assembly requirements of a product, these subdivisions help in specifying the allowable geometrical variation of different surfaces of the product. For example, form needs to be controlled for smooth motion, perpendicularity is important for insertion of long features, and feature size and location must be controlled for proper assembly. ASME Y14.5.1M [15] further elaborates the mathematical description of the concepts in ASME Y14.5M.

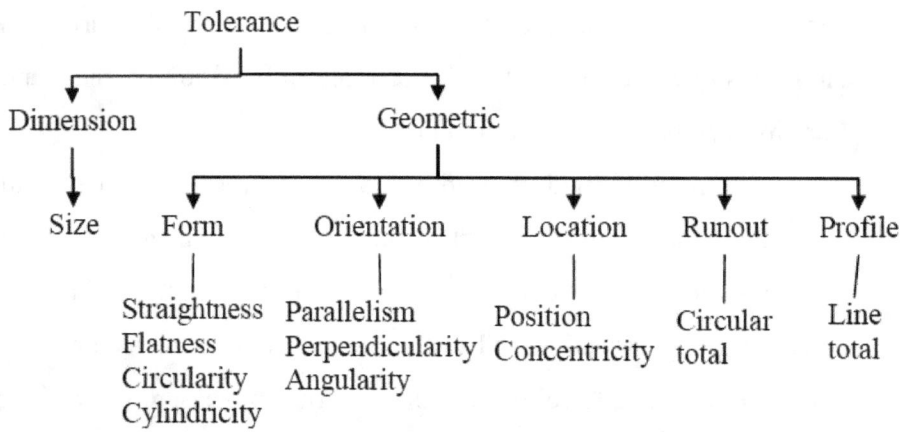

**Figure 3. Classification of tolerance types as defined in the ASME Y14.5 standard**

In the manufacturing stage, there are guidelines for maintaining the quality of the product based on each specific manufacturing process. After a part is produced, adherence to the design specification is measured using different methods, for example functional gages and Coordinate Measuring Machines (CMM). Functional gages are hard gages (go-nogo type) that pass a part if its variation is within the specified tolerance. CMM measurements also provide the amount of variation from the nominal geometry. This feedback to the manufacturing stage, along with other in-process measurements, is provided for monitoring and control of the manufacturing process. All the measurements, either through a go-nogo gage or CMM, have specified uncertainties. To maintain the quality of products despite the uncertainties in measurements and manufacturing processes, statistical and probabilistic tools have been successfully applied. (e. g., statistical process control [16], six-sigma [17]). In the automotive sector, the ISO/TS 16949:2000 specification describes the statistical and probabilistic tools for international quality management. The specification consists of five tools: 1) Advanced Product Quality Planning, 2) Failure Mode and Effects Analysis, 3) Production Part Approval Process, 4) Fundamental Statistical Process Control, and 5) Measurement System Analysis. These five tools provide guidelines to industry best practices and are certified by ISO.

In this section we have described standards related to geometric properties of a product, as an example. We do not intend to discuss all the available standards on product quality with respect to material properties, chemical composition, electromagnetic spectrums, etc. The purpose of studying the standards is to identify requirements that have allowed the successful characterization of the notion of physical quality of products. These requirements are as follows:

a). Standards with clearly defined scope (e.g., ASME Y14.5M)

b). Well classified metrics (e.g., dimensions and geometry)

c). Measurement methods (e.g., functional gages and CMM)

d). Allowable variations in the value of the metric (e.g., tolerances)

e). Application of statistical methods (e.g., statistical process control)

## 3 INFORMATION QUALITY

Physical products, the quality aspect of which was discussed in the previous section, are not the only focus in a manufacturing environment. Physical products are directly interconnected with the information about them throughout the product lifecycle. This product information can

a). represent an input for the manufacturing environment, e.g., information regarding the raw materials to be purchased from the supplier,

b). be generated within the manufacturing environment, e.g., information regarding the product design, or

c). represent an output for the manufacturing environment, e.g., information regarding the maintenance schedule for the product usage.

Physical product is related to product information likewise the quality of the physical product is related to the quality of product information (or information quality, IQ). For example:

a). an enterprise cannot assure the quality of its product if the information on its raw materials is incomplete.

b). an engineer cannot correctly design a product for which the requirements are not properly specified.

c). an enterprise cannot follow a best practice if the information about the best practice is ambiguous.

Some of the principal problems related to IQ are identified by Strong *et al.* [18]. The authors outline IQ problems occurring during information generation (e.g., subjectivity of the information source), storage (e.g., tradeoff between high volumes of stored information and quick access time of the storage) and utilization (e.g., conflict between easy access to information and requirements for security).

This section attempts to extend the notion of quality from physical metrology to information metrology. A similar effort was reported by Ballau *et al.* [19], drawing an analogy between manufacturing systems and information systems, thus enabling a comparison of manufactured product metrology and information product metrology. The approach presented by Lee *et al.* in [5] is part of the Total Data Quality Management (TDQM) framework, which advocates continuous data quality improvement through cycles of define, measure, analyze, and improve [20]. This same framework is adopted by the Department of Defense [21]. As an extension to this framework, Shankaranarayanan and Wang proposed the IP-MAP (information product map) method [22]. The purpose of IP-MAP method is to systematically model the manufacture of the information product. At each stage of information manufacturing, the quality of the information product has to be evaluated.

Several other authors have studied the analogy between manufacturing systems and information systems so as to define some metrics for information quality. These authors define the notion of IQ in terms of: a) its different dimensions and b) how these dimensions affect the end user. In this paper we use the term IQ with a similar connotation, i.e., we define IQ in terms of its dimensions and their impact on the perceived performance of the information system. In our literature survey, the metrics for IQ proposed by various authors can be distinguished, as we mentioned in Section 1, into two broad categories: objective measures and subjective measures. For example, the objective measures could be as trivial as, at the level of data, bits and bytes [10] or as complex as, at the level of semantics, semantic distance and semantic similarity [23]. While objective measures are clearly quantifiable, subjective measures, due to their inherent nature, are hardly quantifiable. For example, believability is hard to quantify and depends on the context. Many authors have attempted to define various subjective metrics for IQ. In the literature, IQ metrics have been referred to as pertaining to elements, aspects, characteristics and properties. In this paper we use only the term information quality dimensions. A brief literature survey of IQ is provided here.

The literature of defining and measuring IQ is primarily developed in the field of computer science and, with some exceptions, in the field of management. Four aspects of information quality are at the center of the research debate: a) identification of the information quality dimensions, b) definition of the dimensions, c) classification of the dimensions and d) metrics for the evaluation of the dimensions. There is no agreement among the authors surveyed on any of these aspects.

Klischewski and Scholl define eight dimensions of IQ: accuracy, objectivity, currency, authority, assurance/reliability, relevance/precision/recall, timeliness and perceived value [24]. A different set of twenty dimensions with their definitions is provided by Lee *et al.* [25]. Although some of the dimensions have the same names, their definitions differ widely. For example, Klischewski and Scholl [24] define timeliness as "an indicator of how fast an information seeker can access the information he/she is looking for," while Lee *et al.* [25] define the same dimension as "the level of information being periodically appropriate to be utilized on executing the user's affairs." Neither author provides any classification of their dimensions.

Other classifications of IQ dimensions have been introduced by Ying and Zhanming [26], Lee *et al.* [5] and Marotta [4]. Ying and Zhanming defined and classified fifteen IQ dimensions in a matrix. Along the rows, IQ dimensions are classified into four categories as Syntactic, Semantic, Pragmatic and Physical based on semiotic theory [26]. The authors define:

a). the syntactic category as the degree to which stored data conform to stored metadata.

b). the semantic category as the degree to which stored data correspond to represented external phenomena.

c). the pragmatic category as the degree to which stored data are suitable and worthwhile for a given use.

d). the physical category as the IQ dimensions that relates to the infrastructure on which the content management process runs and through which the information is actually provided.

Along the columns, they classify the dimensions into definition, assessment, analysis and protection, based on mechanisms of quality assurance. The matrix also provides interconnectivity of the IQ dimensions.

Lee *et al.* provide a survey of academics and practitioners of management of information systems [5]. The authors classify the dimensions mentioned in the survey into four categories:

a). the intrinsic category contains the dimensions representing IQ in its own right

b). the contextual category contains the dimensions representing IQ within the context of the task at hand

c). the representational category emphasizes the importance of computer systems that store information

d). the accessibility category emphasizes the importance of computer systems that provide access to information.

Marotta expands the categories provided by Lee *et al.* [5], where only the system point of view was taken into consideration [4]. The two new categories aim to represent the user point of view:

a). the content category refers to the IQ dimensions relative to the information content perceived by the user

b). the operational category refers to the IQ dimensions that guarantee the access to the information.

Marotta also maps the dimensions belonging to the system point of view into the dimensions belonging to the user point of view [4].

Because of the multiple IQ dimensions and classification systems in the literature, we make an attempt, as an example, to map the IQ dimensions given by Ying and Zhanming [26] to Lee *et al.* [5] (see Figure 4). The mapping is based on our best understanding of the IQ dimensions as provided by the authors. The aim of the mapping is to find similarities and disparities between the IQ dimensions and the classifications suggested by the different authors.

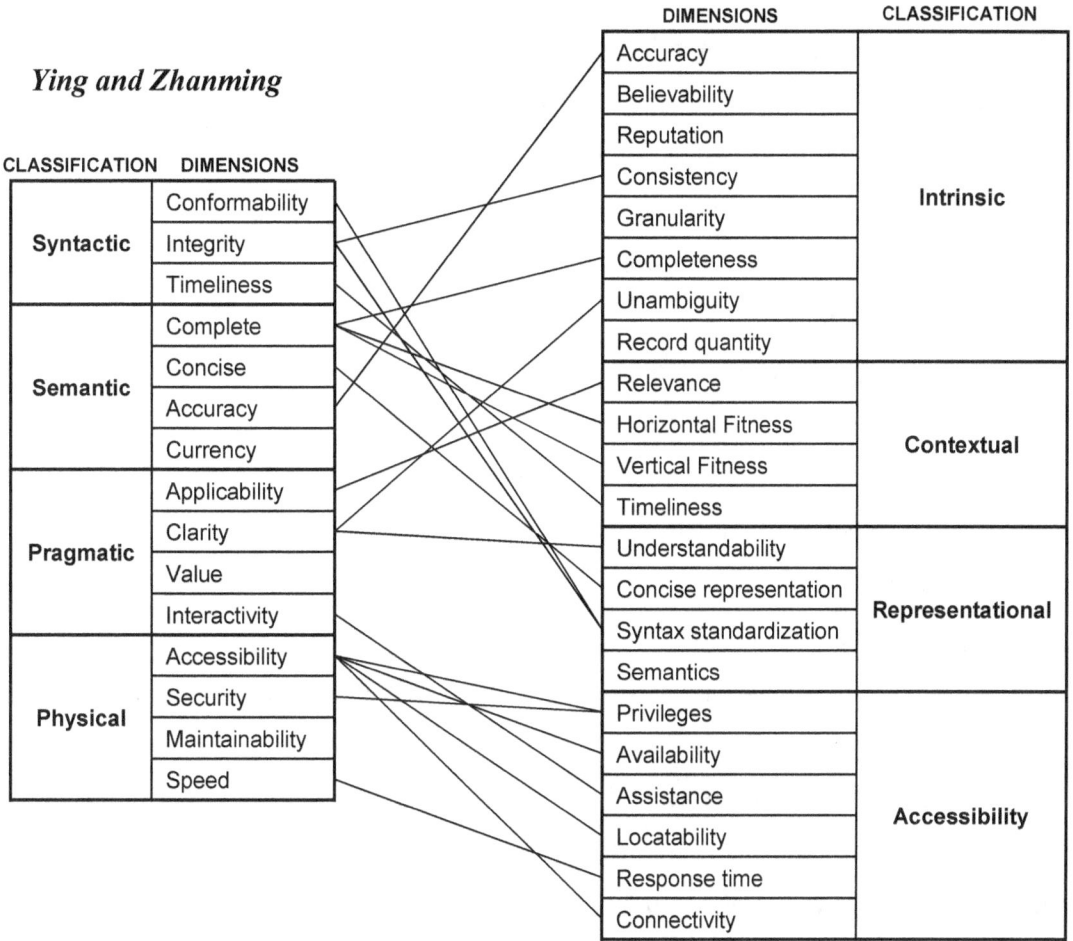

**Figure 4: IQ dimensions mapping ( [26] , [5])**

It is quite clear that many of the dimensions have one to one mapping while others have many to one or none to one. There seems to be no mapping between the classification axes except for the correlations between physical category and accessibility category.

Despite the large effort expended so far, the problem of defining and evaluating the appropriate IQ dimensions is still a persistent research issue. Based on our understanding of the research presented above, there is a lack of harmonized effort towards defining information quality. For defining and measuring information quality we need to put more effort into the following:

a). clear definitions of
   i). information quality dimensions
   ii). classification of the dimensions
b). methods for measuring/evaluating the IQ
   i). how to measure the dimensions of IQ
   ii). how to combine the dimensions for defining IQ metrics
c). allowable variations due to
   i). subjectivity of the information
   ii). information system uncertainties
   iii). IQ evaluation system uncertainties
d). statistical methods utilizing the uncertainties to
   i). improve IQ to find the dimensions that have the greatest impact on IQ and to optimize their value.
   ii). find the economical trade off between available information system and IQ performance.

In the next section we extend the notion of quality from physical metrology to sustainability metrology. We present how the lack of definitions and measuring methods impact not only the information quality of the product but also its sustainability quality. Thus, we complete the assessment of the quality triangle presented in Figure 1.

## 4  SUSTAINABILITY QUALITY

Sustainable development has been defined as "the development that meets the needs of the current generations without compromising the ability of the future generations to meet their own needs" [27]. A definition of sustainability according to the US National Research Council is "is the level of human consumption and activity, which can continue into the foreseeable future, so that the systems that provide goods and services to the humans, persists indefinitely" [28]. Other authors (e.g., Stavins *et al.* [29]) have argued that any definition of sustainability should include dynamic efficiency, should consist of total welfare (accounting for intergenerational equity) and should represent consumption of market and non-market goods and services.

The notion of sustainability has gained worldwide interest due to the current climate change scenario. According to the United Nations Environment Program, climate change is affected by various human activities such as land use changes (through urbanization and deforestation) and fossil fuel burning (through transport, heating, agriculture, industry) [30]. As a result of these human activities, the environment is adversely impacted by not only global warming but also by ozone depletion, acidification, eutrophication, fossil fuel depletion, habitat alteration, air pollution, ecological toxicity, poor human health, and smog formation. Fossil fuel burning and land use changes both influence the carbon cycle and the percentage of $CO_2$, $CH_4$, $N_2O$ and other greenhouse gases. These changes then lead to debilitating environmental effects. For a complete definition of the environmental impacts please refer to [31].

Human activities can be controlled so as to cause least impact on the environment by effectively managing the use/removal of material, energy and waste during a product life cycle. Therefore, for the design, manufacturing, operation, maintenance and recycling of a product, the quality requirements also need to include sustainability. The function of a product also includes its environmental impacts. Figure 5 depicts the idea through a flow diagram. Traditionally, business policy, customer needs, regulations and other requirements have governed the function of a product. Sustainability is now influencing the business policies, customer needs and regulations that influence the function of a product. The dashed arrows in the figure indicate indirect influence while the bold arrows indicate direct influence.

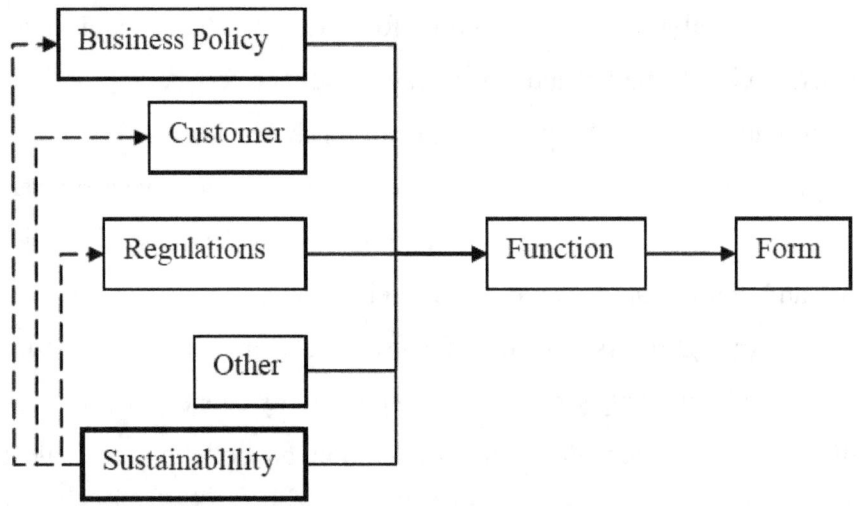

Figure 5. Introduction of the sustainability aspect for deciding the function of a product

A sustainable product can be identified as a product which can be produced, distributed and used in a sustainable manner. A sustainable product can only be developed in a dynamic system working with ever-changing constraints, where inputs and useful outputs are optimized while harmful outputs are minimized. Figure 6 illustrates this concept. A system consists of an enterprise at its core. The enterprise is driven by market, society and environmental requirements. Each enterprise takes input from other enterprises and creates outputs for other enterprises or the consumer. The entire system works under various control parameters ($C$) such as availability of scarce resources, energy efficiency, and governmental and environmental regulations. The system gets resources from the planet as input ($I$) and generates useful and harmful outputs ($O_u$ and $O_h$). There is also a feedback to this system from $O_u$ in terms of reuse, recycle, and remanufacturing. Such a dynamic system can be sustainable if it is resilient and desirable, both within temporal and space scales [32]. We believe that multidisciplinary research in feedback-controlled dynamical systems for sustainability can lead to a better understanding of the interactions among the multiple dimensions of economics, ecology and society.

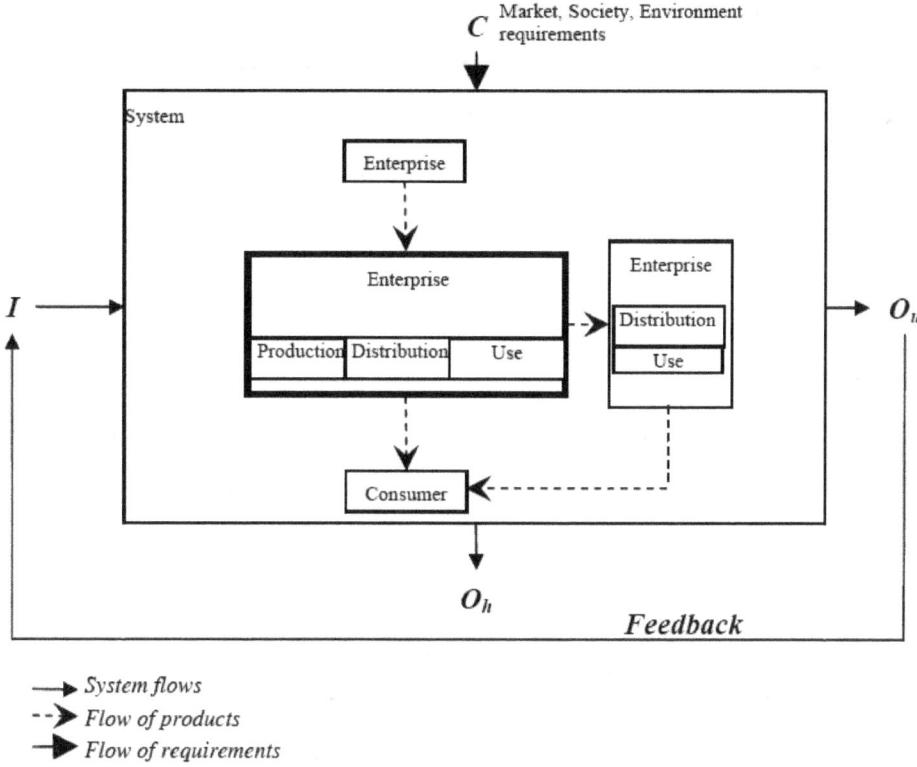

Figure 6. Possible system for producing sustainable products

It is difficult to pinpoint a definition of a sustainable product. We, here, merely attempt to provide the current understanding of the concept of a sustainable product. A real definition of sustainable product can only be obtained through multidisciplinary research.

Datschefski proposed that the quality of a sustainable product be measured using sustainability criteria such as recyclability, safety, efficiency, use of renewable energy and social effects [33]. *Recyclability* implies that the materials used for producing, distributing and using a product can later be useful for some other enterprise in a closed loop. *Safety* indicates not just safety in use to humans but includes the safety in all of releases to air, water, land or space from the production, distribution and use of a product. This would again indicate that all byproducts should be safely consumable in other enterprises/environmental systems. High *efficiency* implies less (than the current) use of energy, material and water during production, distribution and use of a product. *Use of renewable energy* indicates that the product be produced, distributed and used by consuming as much renewable energy as possible in a cyclic and safe manner. *Social effects* refer to the support of basic human rights and natural justice in production, distribution and use of a product.

Labeling a particular product based on its environmental impacts (as given by EPA [31]) requires a mapping of the sustainability criteria to the measures of environmental impacts. Figure 7 demonstrates possible relations that the criteria for sustainability of a product have with measures of environmental impacts. Further multidisciplinary research is required to quantify the transformations (weighting, aggregating, etc.) between the criteria of the sustainability quality of product and the environmental impacts.

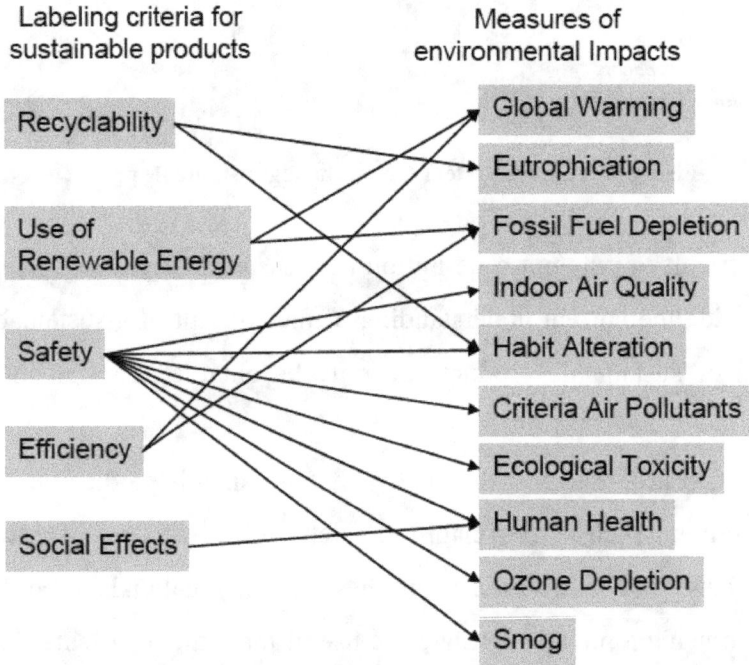

**Figure 7. Potential connections between the criteria for measuring the quality of sustainable products and the environmental impacts [63]**

A product may be of good mechanical quality but may not be of good sustainability quality. One example is the conventional light bulb. It satisfies the function of providing light but it consumes a large amount of electricity. Such a product is being phased out by more efficient and sustainable products (e.g., florescent and LED bulbs).

**Sustainability Standards**

The achievement of sustainability quality of products requires standards that guide the sustainable development of a product. Various standards have been developed in the last two decades to guide sustainable development. A summary of some of these standards is provided in Table 1. The list is not meant to be comprehensive and there is no intent of giving preference to the included standards. These are just a sample of the available standards.

ISO 14000 standards create a systematic approach for reducing the impact on the environment due to the activities of an organization [34]. ISO 14000 standards include the ISO 14020 series for environmental labeling, ISO 14040 for Life Cycle Assessment, ISO 14064 for Green House Gases, to name a few. ISO 19011 provides guidelines for auditing quality and environmental management systems [35].

WEEE, is an acronym for the "Waste Electric and Electronic Equipment" directive [36]. Basically the WEEE directive makes the manufacturers of equipment responsible for the waste. Therefore, the manufacturer should have the infrastructure available to recycle/reuse/process the waste equipment at the end of product's life.

RoHS stands for the "Restriction of Hazardous Substances" directive [37]. It lays down the limit (0.1% by weight) on the use of Lead, Mercury, Cadmium, Hexavalent Chromium, Polybrominated biphenyls and Polybrominated diphenyl ethers, separately, in electronic equipment.

REACH is an acronym for the "Registration, Evaluation, Authorization and Restriction of Chemicals" regulation [38]. It imposes health and safety *evaluation* of all chemicals of one ton or more by *registering* with European Chemicals Agency for *authorization.*

ELV stands for "End of Life of Vehicles" directive [39]. It is similar to WEEE, but is imposed on automotive manufacturers instead on electronics/electrical manufacturers. All electronic equipment in an automobile should follow the ELV directive. IMDS [40], IPC-1752 [41] and JIG-101 [42] are acronyms for "International Material Database System," "Institute of Printed Circuits" and "Joint Industry Guide," respectively. IMDS manages materials for automotive manufacturers, while JIG-101 and IPC-1752 manage material for electronic equipments.

Table 1. Summary of standards for sustainability.

| Standard | Year | Region | Application |
| --- | --- | --- | --- |
| BS 8900 [43] | 2006 | British | managing sustainable development |
| ELV | 2000 | Europe | automotive vehicles |
| Energy Star [44] | 1992 | USA | products, buildings |
| EPA's AP-42 [45] | 1995 | USA | emissions factors for stationary sources |
| IEEE 1680 [46] | 2006 | USA | personal computer products |
| IMDS | 2000 | International | automotive industry material data system |
| IPC 1752 | 2007 | USA | materials declaration in products |
| ISO 14000 | 1992 | International | processes |
| ISO 19011 | 2002 | International | environmental management systems |
| JIG-101 | 2005 | International | materials declaration in products |
| LEED [47] | 1998 | USA | buildings, homes |
| NSF-140 [48] | 2007 | USA | carpet industry |
| REACH | 2006 | Europe | products with hazardous materials. |
| RoHS | 2003 | Europe | new electrical and electronic equipments |
| WEEE | 2002 | Europe | all waste electrical and electronic equip. |

**Sustainability metrics**

The notion of sustainability has received some critical remarks. Scoping sustainability and defining clear system boundaries are critical for properly defining metrics for sustainable manufacturing [49]. Various metrics developed so far to measure the progress towards sustainability have been classified by Mayer [32] and Jain [50] into: a) indicators, b) indices and c) frameworks:

a). *Indicators* basically measure a single parameter of a system, e.g., $CO_2$ emission or energy use. Indicators can be classified into various types such as descriptive, normalized, comparative, structural, intensity, decomposition, causal, consequential, and physical. A detailed survey of indicators has been conducted by Patlitzianas *et al.* [51]. Keffer *et al.* propose a framework for developing a classification of indicators [52]. In the framework, indicators are classified based on aspects and categories. Categories are broad areas of influence related to environment, economy and society, referred to as the triple bottom line of sustainability. Aspects are defined as general type of data that is related to a specific category. Indicators then become the specific measurement of an individual aspect that can be used to demonstrate the status and performance of a system relative to a particular aspect and category.

b). *Indices* are basically aggregates of several indicators, e.g., Ecological Footprint (a ratio of the amount of land and water required to sustain a population to the available land and water for the population) or Environmental Vulnerability Index (consists of indicators of hazards, resistance and damage). Indices represent a single score by combining various indicators of different aspects of a system. Key requirements for sustainability indices, as proposed in Bohringer and Jochem [6], are:

i). Rigorous connection to the definitions of sustainability

ii). Selection of meaningful indicators representing the holistic fields

iii). Reliability and availability of data for quantification over longer time horizons

iv). Process oriented indicators selection

v). Possibility of deriving political objectives

vi). Adequate normalization, aggregation and weighing of the underlying variables

Scientifically sound methods of normalization, weighing and aggregation are a prerequisite for construction of sustainability indices.

The strengths and weakness of several sustainability indices are compared by Mayer [32]. The authors identify several issues across sustainability indices: system boundaries, data inclusion, standardization and weighing methods, aggregation methods, comparisons across indices. Rigorous mathematical requirements for indicators are presented by Ebert and Welsch [53].

c). *Frameworks* present large numbers of indicators in qualitative ways, e.g., the vulnerability framework [54] or the CRITINC Framework [55]. Frameworks do not aggregate data in any manner. An advantage of frameworks is that the values of all indicators can be easily observed and are not hidden behind an aggregated index. The disadvantage of using frameworks is that they are hard to compare over time although this is possible by using Hasse diagrams [56]. A brief review of sustainability frameworks is provided by Mayer [32].

Keffer *et al.* classified indicators into core indicators, that are applicable to all businesses, and supplemental indicators, that are selected based on needs of a particular business and its stakeholders [52]. We believe that this same classification can be applied not only to indicators but, in general, to all metrics, i.e., to indicators, indices and frameworks.

Our literature review shows a considerable proliferation of sustainability metrics that are inconsistently defined and business-specific. Core sustainability metrics, i.e., metrics that are uniformly defined and globally harmonized, are clearly missing. Since core sustainability metrics are the ones that allow for a common definition of sustainability quality, we believe that there exists a vast opportunity for metrology to identify core metrics in the sustainability field.

**Future directions**

Based on our understanding of the research presented above, the notion of sustainability quality clearly lacks definition of a standard terminology, well classified metrics and measurement methods. For characterizing the notion of sustainability quality we identify the following needs

a). *Terminology:* Various terms and definitions used in sustainability should be standardized and harmonized.

b). *Metrics:* There is a pressing need to develop metrics and metrology for various levels of sustainability, viz., (economic) business, (technology) engineering, environment and society level.

c). *Standards:* Due to existence of standards/guidelines in different product categories, different geographical locations etc., harmonization and extension of sustainability standards is desired.

d). *Testing:* With the development of metrics and metrology at different levels of sustainability, testing for compliance with the standards needs to be harmonized.

e). *Information Models:* Extension of current product information models is required to include sustainable development criteria so that such information is available throughout product life cycle.

## 5 DISCUSSION AND COMMENTS

In this paper, we first surveyed quality definitions and identified the problems in defining subjective quality. We then presented our notion of quality as a quality triangle, with its three attributes: physical, information and sustainability.

Second, we showed that the requirements for successfully characterizing the notion of physical quality are clear standards, metrics, measurement methods, allowable variations and statistical methods. We provide examples for each of these requirements.

Then, by using the analogy between manufacturing systems and information systems, we underline the need for research on the requirements of physical quality as applied to information quality. Our survey of the literature on information metrology clearly demonstrated that the scientific community has not yet found an agreement regarding: a) the identification of the information quality dimensions, b) the definition of the dimensions, c) the classification of the dimensions and d) the metrics for the evaluation of the dimensions.

The significance of IQ can be understood from Figure 8, modified from [52]. The figure shows a framework that relates information with the product lifecycle. Data, information and knowledge occupy the inner circle of the framework. Information representation constitutes the second circle. Some of the fundamental activities performed to organize information are included in the middle blue circle: e.g., information modeling, visualization models, and information discovery. The white gap toward the outer blue circle corresponds to the step for modeling complex systems:

products and processes. The information about product and processes is then exchanged between all the phases of the product lifecycle.

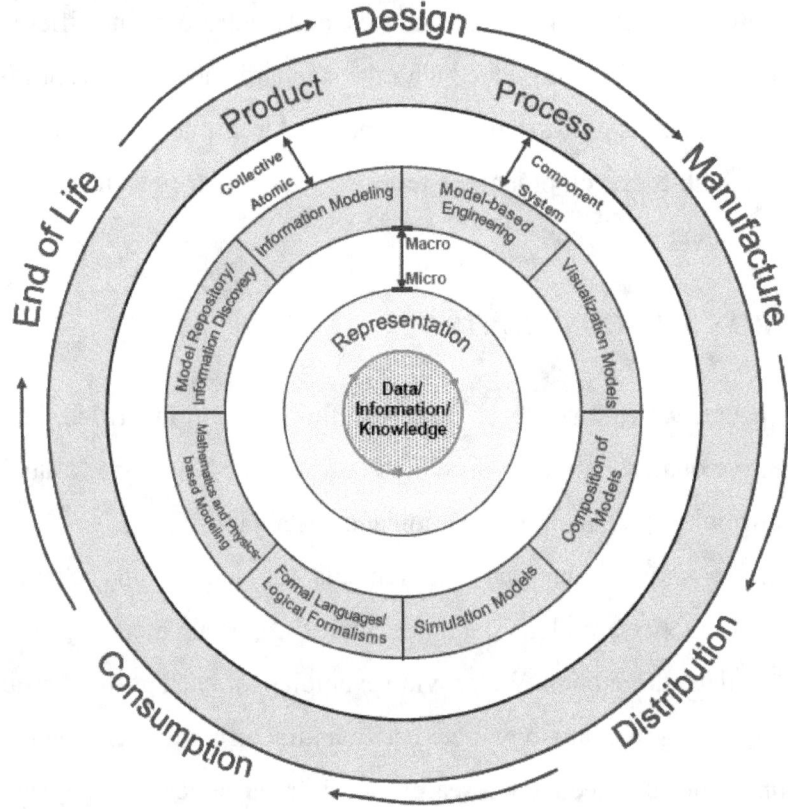

Figure 8. Framework relating information representation and Product Life Cycle (from [52])

IQ plays an important role at each level of the framework. The notion of IQ impacts the whole framework, e.g., from the relevancy of the information for each product lifecycle stage, to expressivity of the modeling formalisms, to readability of the visualization models, to the integrity over the applications transformations.

A lack of IQ at the center of the framework propagates through all the levels and affects the quality of product throughout its lifecycle.

Product quality also includes sustainability quality. In this regard, we introduced several definitions of sustainability. The concept of a sustainable product was presented within a dynamic system consisting of input, useful and harmful outputs and control parameters. We then summarized the current research status on sustainability standards and metrics.

Traditionally, it has been proposed that sustainability can be achieved by managing economic, environmental and societal aspects, called the triple bottom line for sustainability. In a recent article by Sikdar, indicators were identified as 1-D metric as they would quantify changes in only one of the bottom lines of sustainability [7]. Indices could be a 2-D metric or 3-D metric, in a sense that they could quantify changes in either two or three of the bottom lines of sustainability.

In this regard, we propose the triple bottom line for sustainability to be viewed under the technology lens (Figure 9), in order to select the set of technologies with respect to economy, environment and society. Sustainability should not be seen as hindrance to economic growth but as new opportunities to innovate by developing technologies that are both efficient and environmentally benign. However, the challenge is to identify new technologies that would help in quantifying sustainability quality with respect to the triple bottom line.

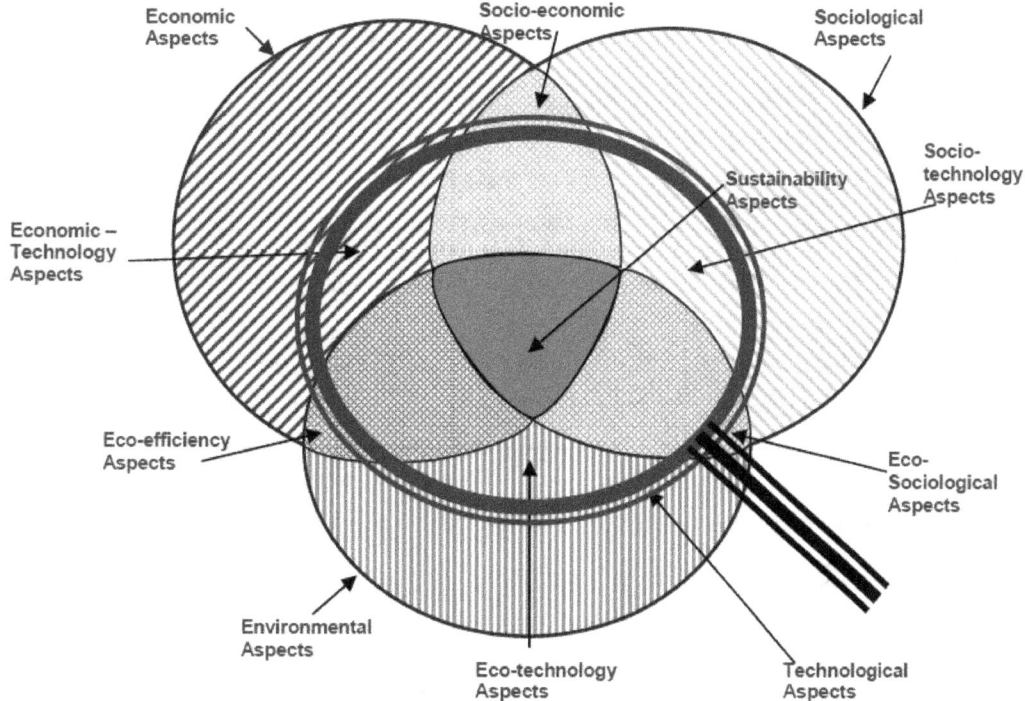

Figure 9. Triple bottom line of sustainability being viewed from a lens of influence of technology

This paper recognizes the need of clear standards, metrics, measurement methods, allowable variations and statistical methods to characterize the notion of product quality. These concepts are utilized to compare physical, information and sustainability qualities as shown in Table 2.

Table 2. Comparison of current state of Quality in Physical, Information and Sustainable realms

|  | **Physical Quality** | **Information Quality** | **Sustainability Quality** |
|---|---|---|---|
| **Standards** | Clearly defined standards with minimal overlaps. | Standard guidelines for data quality exist. Standards on information quality are missing. | Application specific and general guidelines exist. |
| **Definition and classification of metrics** | Metrics defined in generic ways that do not change with time or circumstances. Well classified metrics that represent different functionalities of the product. | Metrics have been defined by different authors, but there is no consensus. | Physical metrics are combined using a common unit to quantify sustainable quality. |
| **Measurement of metrics** | Metric measurable within a given accuracy. | Metrics suggested are mostly subjective. | Various issues with measurement of metrics at the system level. |
| **Allowance for uncertainty** | Allow a range of variation of the metric for unaccountable uncertainties of the system. | Concepts exits but are immature | - |
| **Statistical Methods** | Successful application of Six Sigma and Statistical Process Control. | Concepts exits but are immature | - |

It is evident from the discussion in this paper and the comparison presented in Table 2 that the information and sustainability quality domains lack; a) well scoped out standards, b) a clear metric definition and classification, c) a methodology of measurement of the metric, d) tools for allowing uncertainty and e) probabilistic methods for maintaining quality despite the uncertainty.

In this paper we attempted to include physical, information and sustainability quality into the general notion of product quality. Based on the well-defined concepts in physical metrology, this paper identifies future research direction for achieving measures of information and sustainability quality for products.

# References

1. Hoyer R.W and Hoyer B.Y.B, "What is Quality?," *Quality Progress*, Vol. 34, No. 7, 2001, pp. 53-62.

2. "ISO 9001:2000, Quality Management Systems - Requirements," International Organization for Standardization (ISO), Geneva, Switzerland, 2000.

3. American Society for Quality. http://www.asq.org/glossary/q.html . 2008.

4. Marotta, A.. Quality Management in MSIS. http://citeseer.ist.psu.edu/588629.html . 2007.

5. Lee, Y. W., Strong, D. M., Khan, B. K., and Wang, R., "AIMQ: a methodology for information quality assessment," *Information and Management*, Vol. 40, No. 2, 2002, pp. 133-146.

6. Bohringer C.and Jochem P.E.P., "Measuring the immeasurable - A Survey of sustainability indices," *Ecological Economics*, Vol. 63, 2007, pp. 1-8.

7. Sikdar S.K, "Sustainable Development and Sustainability Metrics," *AIChE Journal*, Vol. 49, No. 8, 2008, pp. 1928-1932.

8. Swyt, D. A., "An Assessment of the United States Measurement System: Addressing Measurement Barriers to Accelerate Innovation," National Institute of Standards and Technology, Gaithersburg, NISTIR 6025, 2007.

9. World Metrology Day 2004: Summary. http://www.bipm.org/en/convention/wmd/2004/ . 2004.

10. Carnahan, L., Carver, G., Ray, M., Hogan, M., Hopp, T., Horlick, J., Lyon, G., and Messina, E., "Metrology for Information Technology (IT)," National Institute of Standards and Technology, Gaithersburg, NISTIR 6025, 2007.

11. "ISO/IEC Guide 99:2007 International Vocabulary of Metrology - Basic and General Concepts and Asoociated Terms (VIM)," International Standards Organization, Geneva, Switzerland, 2007.

12. "ISO/IEC Guide 98:1995 Guide to the Expression of Uncertainty in Measurement, GUM," International Standards Organization, Geneva, Switzerland, 1995.

13. "ISO 1101:2004 Geometrical Product Specifications (GPS) -- Geometrical tolerancing -- Tolerances of form, orientation, location and run-out," International Organization for Standardization, Geneva, Switzerland, 2004.

14. "Dimensioning and Tolerancing," American Society of Mechanical Engineers, New York, ASME Y14.5M-1994, 1994.

15. "Mathematical Definition of Dimensioning and Tolerancing Principles," ASME Y14.5.1M-1994, 1994.

16. Wetherill G.B.and Brown D.W., *"Statistical process control: theory and practice"*, Chapman and Hall.1991.

17. Harry M.J., *"The Nature of Six Sigma Quality"*, Motorola University Press.1988.

18. Strong, D. M., Lee, Y. W., and Wang, R., "10 Potholes in the road of information quality," *Computer*, Vol. 30, No. 8, 1997, pp. 38-46.

19. Ballau, D., Wang, R., Pazer, H., and Tayi, G. K., "Modeling Information Manufacturing Systems to Determine Information Product Quality," *Management Science*, Vol. 44, No. 4, 1998, pp. 462-484.

20. Madnick, S. and Wang, R., "Introduction to the TDQM Research Program," *TDQM Working Paper Series*, 1992.

21. DOD Guidelines on data quality management. http://www.tricare.mil/ocfo/_docs/DoDGuidelinesOnDataQualityManagement.pdf . 1996.

22. Shankaranarayanan, G., and Wang, R., "IPMAP: Research Status and Direction," Proceedings of the 12th International Conference on Information Quality,2008, pp. 510-517.

23. Resnik, P., "Semantic Similarity in a Taxonomy: An Information-Based Measure and its Application to Problems of Ambiguity in Natural Language," *Journal of Artificial Intelligence Research*, Vol. 11, 1999, pp. 95-130.

24. Klischewski, R., and Scholl, H. J., "Information Quality as a Common Ground for Key Players in e-Government Integration and Interoperability," Vol. 4, Proceedings of the 39th Annual Hawaii International Conference on System Sciences,2006.

25. Lee, J., Lee, Y., Ryu, Y., and Kang, T., "Information Quality Drivers of KMS," Proceedings of the 2007 International Conference on Convergence Information Technology,2007, pp. 1494-1499.

26. Ying, S., and Zhanming, J., "Assuring Information Quality in Knowledge Intensive Business Services," International Conference on Wireless Communications, Networking and Mobile Computing,2007, pp. 3243-3246.

27. *Our Common Future*, Oxford University Press,1987.

28. *Our Common Journey: A Transition Toward Sustainability.*, National Academy Press 1999.

29. Stavins R.N., Wagner A.F., and Wagner G., "Interpreting sustainability in Economic Terms: Dynamic Efficiency Plus Intergenerational Equity," *Economic Letters*, Vol. 79, No. 3, 2003, pp. 339-343.

30. Rekacewicz, P., "Climate change: processes, characteristics and threats," UNEP/Grid-Arendal Maps and Graphics Library, 2005.

31. Bare, J. C., Norris, G. A., Pennington, W., and McKone, T., "TRACI – The Tool for the Reduction and Assessment of Chemical and Other Environmental Impacts," *Journal of Industrial Ecology*, Vol. 6, No. 3, 2002, pp. 49-78.

32. Mayer A.L., "Strengths and weakness of common sustainability indices for multidimensional systems," *Environment International*, Vol. 34, No. 2, 2008, pp. 277-291.

33. Datschefski E., *The Total Beauty Of Sustainable Products*, Rotovision SA, Switzerland, 2001.

34. "ISO 14001:2004, Environmental management systems -- Requirements with guidance for use," ISO, International Organization for Standardization (ISO), Geneva, Switzerland, 2004.

35. "ISO 19011:2002, Guidelines for quality and/or environmental management systems auditing," International Organization for Standardization (ISO), Geneva, Switzerland, 2002.

36. "Directive 2002/96/EC of the European Parliament and of the Council of 27 January 2003, on Waste Electrical and Electronic Equipment, Official Journal of the European Union, L37/24-L37/38," 2003.

37. "Directive 2002/95/EC of the European Parliament and of the Council of 27 January 2003, on the restriction of the use of certain hazardous substances in electrical and electronic equipment., Official Journal of the European Union, L37/19-L37/23," 2003.

38. "Regulation (EC) No 1907/2006 of the European Parliament and of the Council of 18 December 2006 concerning the Registration, Evaluation, Authorisation and Restriction of Chemicals (REACH)," 2006.

39. "ELV Directive 2000/53/EC of the European Parliament and of the Council of 18 September 2000 on end-of life vehicles," 2000.

40. MDSYSTEM. http://www.mdsystem.com/index.jsp . 2008.

41. IPC.ORG. http://members.ipc.org/committee/drafts/2-18_d_MaterialsDeclarationRequest.asp . 2008.

42. JEDEC.ORG. http://www.jedec.org/download/search/ACF276.pdf . 2008.

43. "BS 8900:2006 Guidance for managing sustainable development," (ISBN 0 580 481 43 3), 2008.

44. ENERGYSTAR. http://www.energystar.gov/ . 2008.

45. EPA.GOV. http://www.epa.gov/ttn/chief/ap42/ . 2008.

46. "IEEE 1680-2006 Standard for Environmental Assessment of Personal Computer Products, Including Laptop & Desktop Computers & Monitors.," 2006.

47. USGBC.ORG. http://www.usgbc.org/ . 2008.

48. "Draft Standard NSF 140 – 2005: Sustainable Carpet Assessment Draft Standard," NSF International, Michigan, USA, 2005.

49. "Sustainability risks being about everything and therefore, in the end, about nothing," The Economist, 2002.

50. Jain, R., "Sustainability: metrics, specific indicators and preference index," *Clean Technologies and Environmental Policy*, Vol. 7, 2005, pp. 71-72.

51. Patlitzianas, K. D., Doukas, H., Kagiannas, A. G., and Psarras, J., "Sustainable energy policy indicators: Review and recommendations," *Renewable Energy*, Vol. 33, No. 5, 2008, pp. 966-973.

52. Keffer, C., Shimp, R., and Lehni, M., "Eco-efficiency Indicators & Reporting, Report on the Status of the Project's Work in Progress and Guideline for Pilot Application," WBCSD, Working Group on Eco-efficiency Metrics and Reporting, Geneva, Switzerland, 1999.

53. Ebert U. and Welsch H, "Meaningful Environmental Indices: A Social Choice Approach," *Journal of Environmental Economics and Management*, Vol. 47, 2004, pp. 270-283.

54. Turner, B. I., Kasperson, R. E., Matson, P. A., McCarthy, J. J., Corell, R. W., and Christensen, L., "A framework for vulnerability analysis in sustainability science," Vol. 100, USA, 2003, pp. 8074-8083.

55. Ekins, P., Simon, S., Deutsch, L., Folke, C., and De Groot, R., "A framework for the practical application of the concepts of critical natural capital and strong sustainability," *Ecological Economics*, Vol. 44, No. 165, 2003, pp. 185.

56. Patil G.P. and Taillie C., "Multiple indicators, partially ordered sets, and linear extensions: multi-criterion ranking and prioritization," *Environmental and Ecological Statistics*, Vol. 11, 2008, pp. 199-228.

www.ingramcontent.com/pod-product-compliance
Lightning Source LLC
Chambersburg PA
CBHW081808170526
45167CB00008B/3380